T0395086

Slithering Snakes

King Cobra

Snake Eater

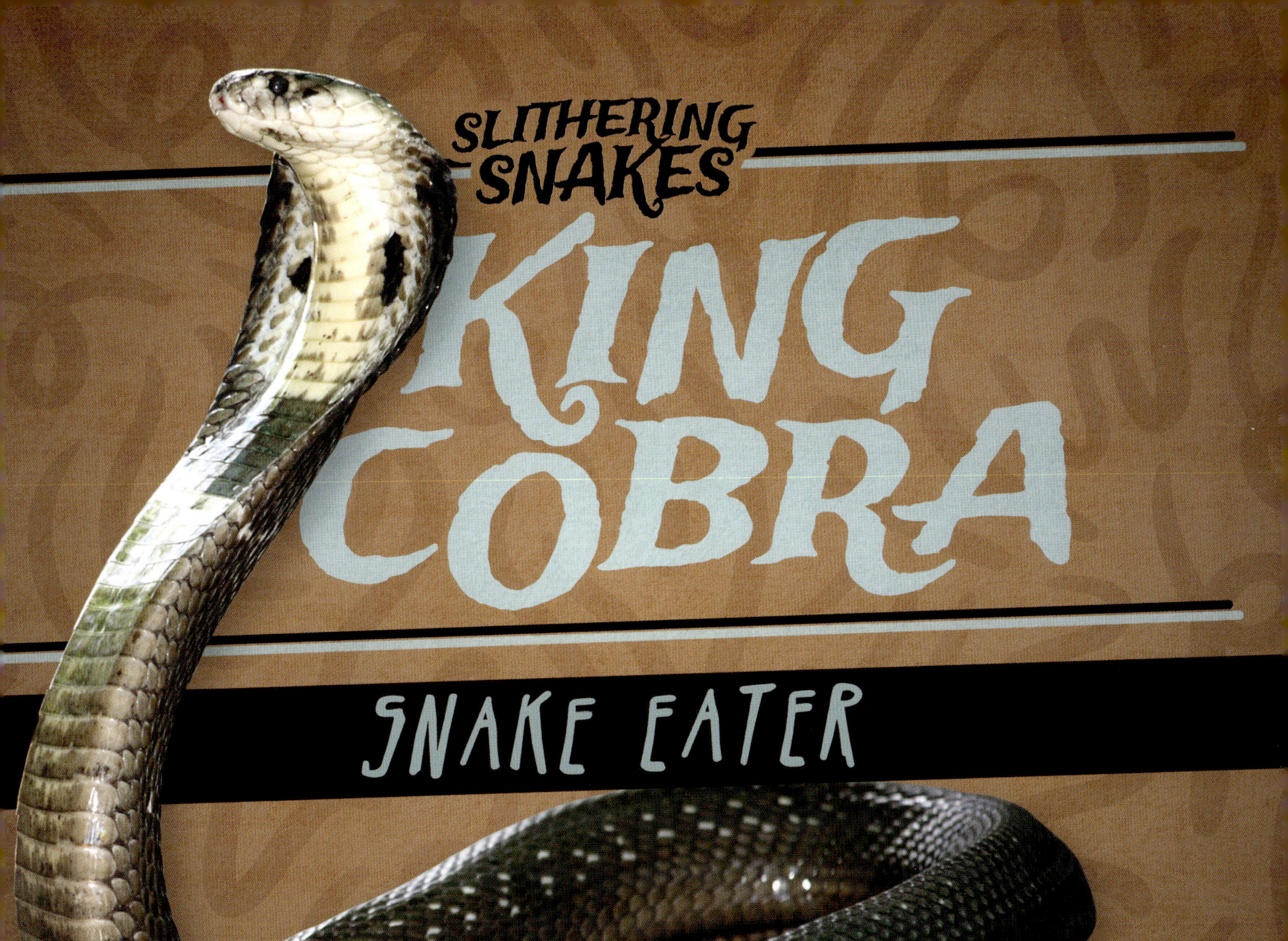

By Natalie Humphrey

Enslow Publishing

DISCOVER!

Please visit our website, www.enslow.com. For a free color catalog of all our high-quality books, call toll free 1-800-398-2504 or fax 1-877-980-4454.

Library of Congress Cataloging-in-Publication Data

Names: Humphrey, Natalie, author.
Title: King cobra : snake eater / Natalie Humphrey.
Description: New York : Enslow Publishing, [2021] | Series: Slithering snakes | Includes index.
Identifiers: LCCN 2020005082 | ISBN 9781978517813 (library binding) | ISBN 9781978517790 (paperback) | ISBN 9781978517806 (six-pack) | ISBN 9781978517820 (ebook)
Subjects: LCSH: King cobra—Juvenile literature.
Classification: LCC QL666.O64 H865 2021 | DDC 597.96/42—dc23
LC record available at https://lccn.loc.gov/2020005082

Published in 2021 by
Enslow Publishing
101 West 23rd Street, Suite #240
New York, NY 10011

Designer: Sarah Liddell
Editor: Natalie Humphrey

Photo credits: Cover, p. 1 (king cobra) potowizard/Shutterstock.com; background pattern used throughout Ksusha Dusmikeeva/Shutterstock.com; background texture used throughout Lukasz Szwaj/Shutterstock.com; p. 5 LenSoMy/iStock/Getty Images Plus/Getty Images; p. 7 Mark Newman/The Image Bank/Getty Images; p. 9 R. Andrew Odum/Oxford Scientific/Getty Images Plus/Getty Images; p. 11 reptiles4all/iStock/Getty Images Plus/Getty Images; p. 13 oariff/iStock/Getty Images Plus/Getty Images; p. 15 ccarbill/Shutterstock.com; p. 17 Image © Black Stallion Photography/Moment Open/Getty Images; p. 19 Arun RaJpuT6621/Shutterstock.com; p. 21okili77/Shutterstock.com.

Portions of this work were originally authored by Audry Graham and published as *King Cobra*. All new material this edition authored by Natalie Humphrey.

Printed in the United States of America

CPSIA compliance information: Batch #BS20ENS: For further information contact Enslow Publishing, New York, New York, at 1-800-398-2504.

CONTENTS

Boldface words appear in Words to Know.

KING OF THE COBRAS

With enough **venom** to kill an elephant, it's no wonder this cobra is called the king! The king cobra can grow up to 18 feet (5.5 m) long and weigh up to 20 pounds (9 kg). This giant snake is the largest venomous snake in the world!

KING
COBRA

KING COBRAS AT HOME

King cobras spend most of their time in rain forests and near streams. They live in northern India and in parts of Southeast Asia. King cobras can be light brown, dark brown, or greenish-brown with yellow or white bellies. They usually have white, brown, or yellow stripes.

KING COBRAS ARE GREAT SWIMMERS.

IN THE NEST

Female king cobras are the only snakes that build nests. Between January and April, a female pushes together branches and leaves for her eggs. She will lay 21 to 40 eggs and keep them safe until they **hatch**. Each baby is about 20 inches (51 cm) long.

KING COBRAS LIVE ABOUT
20 YEARS IN THE WILD.

STAY AWAY!

When a king cobra is scared or angry, it tries to make itself look big and scary. The king cobra uses bones and **muscles** to spread its neck out as wide as it can. This stretching creates the snake's famous hood.

A KING COBRA'S HISS SOUNDS SIMILAR TO A GROWLING DOG.

A scared king cobra can lift its head up to 4 feet (1.2 m) off the ground. It can follow an enemy in this pose for a long time. When it's ready to strike, it brings its head down quickly and drives its **fangs** into its target.

A KING COBRA'S FANGS ARE ONLY HALF AN INCH (1.3 CM) LONG.

KING COBRA VENOM

Unlike many snakes, what makes a king cobra deadly isn't the strength of its venom. King cobra venom isn't very strong. But, king cobra bites have a lot of venom in them. This makes the king cobra one of the deadliest snakes in the world.

DOCTORS STUDIED KING COBRA VENOM TO MAKE SOME **MEDICINES**.

ON THE HUNT

King cobras are picky eaters. When they're hungry, they only want one thing—other snakes! King cobras are known to eat pythons, rat snakes, Indian cobras, and even smaller king cobras. Some king cobras like one kind of snake and will eat only that.

KING COBRAS CAN BE TRAINED TO EAT MICE OR RATS IN ZOOS.

King cobras can see **prey** moving from 330 feet (100 m) away! When a king cobra spots its prey, it **attacks** quickly. It bites the prey and uses its venom to **stun** and kill the animal. Then, the king cobra eats its prey whole.

KING COBRAS DON'T OPEN THEIR HOOD WHEN THEY'RE HUNTING.

KING COBRAS AND PEOPLE

Only a female king cobra keeping her eggs safe will attack anything that moves. Otherwise, king cobras try to stay far away from people. If a human comes too close, king cobras will hiss and try to scare the human away, not attack.

WHERE DO KING COBRAS LIVE?

ASIA

PACIFIC OCEAN

INDIAN OCEAN

WHERE KING COBRAS LIVE

WORDS TO KNOW

attack To try to harm someone or something.

fang A long, sharp tooth.

hatch To break out of an egg.

medicine Matter that is used to treat illnesses or help with pain.

muscle A body part that can produce movement.

prey An animal that is hunted or killed by another animal for food.

stun To shock something so it can't move.

venom Poison that is made by an animal and used to hurt or kill another animal.

FOR MORE INFORMATION

BOOKS

Hamilton, S. L. *Cobras*. Minneapolis, MN: Abdo Publishing, 2018.

Sprott, Gary. *King Cobra*. North Mankato, MN: Rourke Educational Media, 2019.

WEBSITES

National Geographic Kids
kids.nationalgeographic.com/animals/reptiles/king-cobra/
Discover more facts and watch videos about king cobras.

San Diego Zoo
zoo.sandiegozoo.org/animals/king-cobra/
Learn more quick facts about king cobras.

INDEX